The Source

by

Igor Kryan, MS

2006

Contents:

Popular astrobiology research in 3 parts: I. Cosmic Time Bomb. II. The Source. III. Silentio Universum.

*02.2004. Originally written by Igor Kryan, MS as an astrobiology research
for the Amsterdam Ravenhurst University, Netherlands.
01.2006. Modified by the author to fit for the most readers.*

All I am offering is a scientifically based truth - nothing more.
This book will forever change your mind about many things.
Read and discover the real answers for the following questions:

What are the Earth nearest and distant future preparing for us?
What causing of asthma, influenza and some other illnesses?
How and when will the Earth magnetic poles flip occur?
What creating such powerful hurricanes like Katrina?
What does the nearest future prepare for us?
Where did life in the Universe come from?
What scientists do not want you to know?
Where did life on the Earth come from?
How is the planetary climate changing?
Why is the global warming happening?
Why did Viking Era warming occur?
Why should you be avoiding rains?
How did our solar system appear?
When does the doomsday come?
What is going on with our Sun?
Why did Neanderthals extinct?
What is the comic time bomb?
Was there any life on Mars?
Why did dinosaurs die out?
Why the Universe is silent?
What makes life possible?
Is asteroids threat real?
Why did Ice Age occur?
What are UFOs?
Are we alone?

Part One: Cosmic Time Bomb.

In September 2003 scientists spotted asteroid "2003 QQ47" and its first measurements show a considerable probability of a direct hit to the Earth on March 21, 2014 by that huge cosmic rock with an explosion the size of 20 million Hiroshima atomic bombs. Most of the life on the Earth would be obliterated in case of such a catastrophic event. Fortunately, the forecast was revised - it would miss the Earth, after all.

To discover potentially dangerous objects near our planet and to avoid such scares like "2003 QQ47" was, NASA established Near Earth Object Observation Program in 1997 to find everything bigger than 1 kilometer in diameter that might approach the planet. This search started in 1997 and to date more than 800 objects of the total estimated population of over 1,100 have been discovered. Thus, now almost 80 percent of such flying rocks has been found and mapped. All of them are believed to be traced by 2009 or even earlier. Guess how many of found asteroids larger than one kilometer carry even a slightest potential probability of hitting the Earth within the next few centuries? Zero.

According to Near Earth Object Observation Program statistical research, the probability of the Earth being hit by a large object in this century is less then 0.000001. In reality it is even much lower chance because, as most scientists believe, that an event of 1 kilometer asteroid impact would only occur about every 700,000 years on average and considering the fact that humanity already survived one such large scale and rare impact last century. I mean the Tunguska event in 1908. This asteroid or comet (scientists still arguing what it was) flattened 800 square miles of Russian Siberian forest while this object did not even reach the ground level - it was blown apart in the air. The Tunguska could kill millions in case it hit not Siberia but Europe or any other densely populated region. However, no other Tunguska-like event is visible in our nearest or even more distant future.

We should not completely refuse the fact that asteroids may represent a grave threat to life on the Earth - large objects certainly have hit our planet many times. Just an asteroid scare is a simple media exaggeration because the impact is not going to happen in this or next century.

An asteroid or a comet that believed to have kicked up much dust and ignite many volcanoes 65 million years ago, resulting climate changes, wiped out the dinosaurs. However, this assumption is also incorrect. The large cosmic object obliterated dinosaurs 65 million years ago not simply by changing planetary climate and causing years of dusty dark sky and cool weather but also by causing too many dinosaurs to be born males. British scientists David Miller and Sherman Silber have proven this fact several years ago while for some unknown reason it was ignored by the media: "If dinosaurs were like modern-day reptiles such as crocodiles, they change sex based on temperature. And even a small skewing of populations toward males would have led to eventual extinction.

The Earth did not become so toxic that life died out 65 million years ago; the temperature just changed, and these great beasts had not evolved a genetic mechanism (like our Y chromosome) to cope with that." As you may know, for almost all mammals and birds if an embryo gets an X and a Y chromosome, it will be born male and if it gets both X chromosomes it will be born female. But reptiles and some fishes have the temperature at which incubated eggs can affect the sex of the developing babies. However, in some reptiles (like modern day crocodilians and turtles and some fishes), the temperature at which eggs are incubated can also affect the sex of the developing babies. These scientists made an analysis, which clearly shows a temperature shift led to a preponderance of males. Other scientific observations have also shown that when there are very few females in a population of species dies out. The only reptiles which are affected less by this event are crocodiles and turtles, that live at mostly aquatic environment. And water offered some sufficient protection against most effects of environmental changes and gave them more time to became adapt to a cooler climate.

As we can see, that fearsome asteroid or even several asteroids did not actually kill all dinosaurs, in fact, the impact itself probably wiped out less then one percent of their entire population. It only triggered a series of global climatic and temperature changes, which had been occurring for many centuries and, finally, erased the dinosaurs and many other species from the surface of our planet.

The other question "where and why does these asteroid or asteroids came from?" is also had been answered by the group of US and French scientists and also ignored by the media their "Cosmic Time Bomb Theory:" "Jupiter orbits the Sun about five times for every two times Saturn goes round. If the ratio of the orbital periods was precisely 5:2, the combined effect of the gravity of two massive planets on other bodies in the solar system would be greatest every 10 years - that is, when the two planets are on the same side of the Sun and pulling together. But, because this 5:2 resonance is not exact, the planets are in perfect alignment on the same side of the Sun only every 1,000 to 2,000 years." Taking this effect into an account, scientists have discovered that "as the semi-major axis of Saturn's orbit changes, the Jupiter-Saturn system drifts back and forth between motion which is regular and motion which is totally chaotic. The system trips over into chaos every few tens of millions of years... The most remarkable discovery is that in a wide range of simulations in which the semi-major axis of Saturn is allowed to vary, a burst of chaos arises around 65 million years before the present (back to the dinosaurs extinction)... Now it is possible to investigate the effect it would have had on other bodies in the solar system - specifically, asteroids. The asteroids are thought to be the left-over rubble of a planet which was prevented from congealing out of the proto-planetary space by the disruptive effect of Jupiter."

As you may know, a considerable number of asteroids circles the Sun between the orbits of Jupiter and Mars and scientist simulated the effect on asteroids of a burst of chaos in the Jupiter-Saturn system. And the most important effect are abrupt changes in the axis of asteroid orbit which lead eventually to complete ejection of rocky bodies from the asteroid belt and some of them, eventually, right into orbits which cross the Earth's orbit. This had happened 65 millions years ago (dinosaurs extinction) and another asteroids ejection occurred about 250 million years ago (mass extinction at the Permian-Triassic ages boundary). This event happened in the past many times and will happen again, although not always ejected asteroids aimed at the Earth. The due of the next ejection is as far as 30 million AD, so there is plenty of time to get prepared to a new possible Armageddon. In my opinion human kind will think of something for 30 million years, and if not, most likely, huge cosmic bodies will miss us this time, because they struck the planet last time and the statistical probability of two correlated strikes in a raw is only a few points of a percent. Thus, now we can more or less relax for another 30 million years.

Many astronomers as well as Cosmic Time Bomb theory designers are trying to represent our Solar system as a pretty stable and closed system, which may be disrupted once per hundreds of millions years or so. Some scientists of them are even trying to demonstrate that the Earth itself is a closed system limited with a stratosphere by stressing that climate changes, global warming, and other effects are happing only because of human activity. In fact, the first scientist who proved 60 years ago that our Earth is an open to the outer space influence was Russian biologist D. Vernadsky, while absolute majority of modern scientists still not ready to accept the fact that something except a huge asteroid or a comet can lead to tremendous transformations of our planet, such as dinosaurs extinction or ice age.

Michael Ghil, Ferenc Varadi, and Bruce Runnegar, who are main Cosmic Time Bomb theory designers, may be correct - there might be or might not be asteroids collisions 65 and 250 million years ago (nobody still knows for sure if a huge cosmic rock hit Yucatan peninsula 65 million BC or not). Probably, a collision with a cosmic object indeed took its place at that time, and it was just a coincidence but not the main reason why nearly 80% of life on Earth died out. Something less noticeable but much bigger has happened and it will happen again soon.

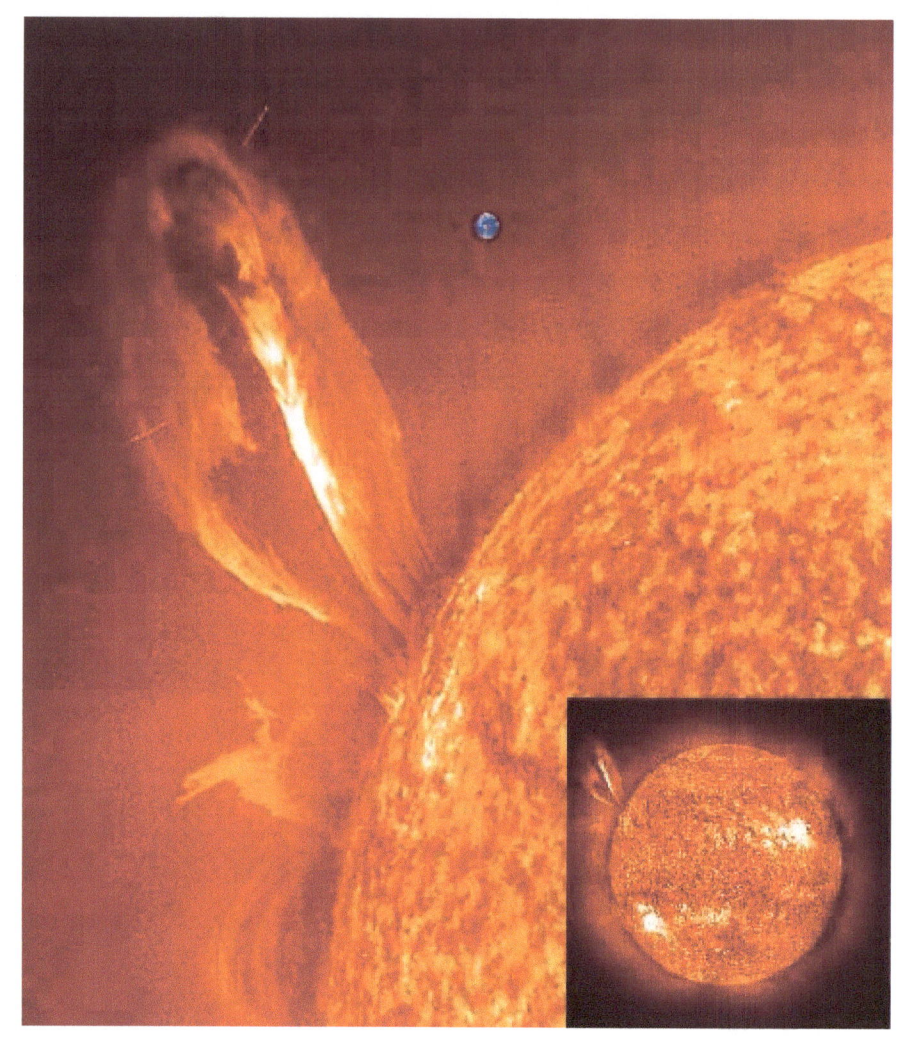

Part Two: The Source.

As we all know, the Sun is our nearest star. Thermonuclear reaction deep inside the Sun creates the light and heat the life on Earth needs for our survival. Our Sun was born about five billion years ago. The Sun is consuming five million tones of hydrogen every second, but it believed to be large enough to continue to shine for another 5 billion years or so. By that time, it will become a red giant, causing the surrounding planets to boil away. Although some optimistically oriented scientists believe that the dying Sun gravity will pull our Earth away from it, so it will give to the planet an additional billion years to exist. As you can see, the Solar system seems to be pretty stable till 5 billion AD. From the first look.

In fact, the energy released by the Sun is not always constant. Every 11 years or so changes in Sun's magnetic field bring on an increase in sunspots, which are known as solar flares - events of coronal mass ejections. They result is a barrage of charged particles hurling toward Earth. The peak in this 11 years cycle is called a "solar maximum." Solar flares, one type of "space weather" associated with solar maximums, release colossal amounts of energy, equivalent to the millions hundred-megaton nuclear explosions in just a few seconds.

However, the Sun, clearly, has some other much longer cycles of its activity. Minimum of solar activity when no solar flares (no dark spots on the surface of the Sun) were visible during the 17th century occurred at exactly the same time as the Little Ice Age, when the average planetary temperature dropped -1 C degree for just a few years. The next minimum solar activity period occurred 1800-1830 coincided with another cool climatic period dubbed "Dickens Winters" (Charles Dickens wrote novels describing snowy Christmases and frozen Thames in London, which normally does not much snow or frozen rivers in winter). 1816 called "The Year Without a Summer", due to the unusual cold which swept in America and Europe that year.

The similar long solar minimum, probably, occurred 65 millions years ago and caused the cooler temperature at which incubated dinosaurs eggs affected the sex of the developing babies and led to a preponderance of males. And, as we learned in the part one, when there are very few females a population of species dies out. That is why dinosaurs are not running on streets today, and that is why we can leave huge fireball collision nightmarish pictures to Hollywood.

Probably, somewhat similar but more serious minimum of solar activity took place about 250 million years ago (mass extinction at the Permian-Triassic ages boundary), when over 80% of primitive species disappeared from the Earth surface.

Not only mentioned dinosaurs but a whole second humanity - Neanderthals became extinct because of the very long Solar minimum well know as the Great Ice Age.

Neanderthals were well developed primitive society who mourned their dead, hunted and gathered in organized way, built caves, created simple tools, and even set headstones. They were five to six times stronger than an average human and much well designed to hunt and live in a wild nature than humans.

The only reason they died out like dinosaurs and lost the Earth to weak humans is because of another solar minimum. When the cooler climate struck, humans invented clothes from animal skins and furs in order to survive and keep themselves warm. And unfortunately (or fortunately for humans who get rid from their competitors) Neanderthals did not develop any language, while humans have already had primitive language system and sign system. In contrast to humans, even if some Neanderthals were also smart enough to create clothes from animal skins and furs, they could not posses the knowledge how to warm themselves to the others, because of absence of any stuffiest communication system.

Don't worry: we will not be frozen into ice and we have a sufficient communication system (at least comparative to Neanderthals). However, something opposite and much more serious (not understood and unavoidable) is occurring now.

All astronomers agree: the Sun is more active now than it has been for a millennium. This conclusion comes from a reconstruction of sunspots (flares mostly occurring during solar maximums). Sunspot observations stretch back to beginning of the 17th century, when the telescope was invented. To extend the age of possible observations farther back in time, scientists used a physical model to calculate past sunspot numbers from levels of a radioactive isotope preserved in ice levels taken from Greenland and Antarctica.

Since "Dickens Winters" the Sun activity and sunspots numbers have been growing constantly and steadily, as well as a temperature on the Earth. It cannot have any coincidence that planet climate warmed up at exactly the same time, as the Sun activity grows. As you can see, green house gases production and burning fossil fuels, that suggest human-induced climate change has a little to do with the global warming. Probably, as soon as US George W Bush administration realized these facts, despite international criticism they dropped Kyoto Protocol, which limits emission of green house gases like carbon dioxide per country.

Climatologists may disagree with me and make a statement that the late 19th century was another cooler period despite the rising activity of the Sun. However, the most powerful (after Tabora) ever recorded on the Earth volcano eruption occurred in 1883. The Krakatoa volcano produced so much ash and thrown so much tiny particles into our atmosphere that amount of the sunlight penetrating Earth atmosphere was reduced by over 5%, which let to the whole planet climate to became cooler. By the way, because of those volcanic tiny particles flying in our atmosphere, the number of people suffering from asthma and other lung illnesses increased over five times the late 19th century. Before 1883 asthma was a very rare and almost unknown illness.

More sunspots are linked to increased discharges of solar energy from the Sun and provide major impact on Earth's climate. Thus, the research from the University of New England in Australia demonstrates that the 500-year cycle is a part of a stronger 1500-year pattern of solar activity which has been linked to historic events such as ice ages and ancient global warming, which occurred during Viking Era (1100 - 1200 AD, when Viking for sure where not burning fossil fuels and producing other greenhouse gases). By the way, during an ancient global warming peak in 1200 AD, when Greenland Island became indeed green not but not icy like now, the average temperature on the Earth was over 0.5 C degree more then now. This period is called Medieval maximum, after that the Sun activity went down with its minimum in 1700 AD (Well known as a Little Ace Age). Since this Little Ice Age the Sun activity has been constantly increasing as a part of the cycle.

These evidences found in the most distant past as well. In September 2004, Swedish-flagged, Norwegian-operated drillship ventured to within 238 kilometers close to the North Pole, protected by a Russian icebreaker, which broke up sea ice around the vessel as it drilled into the ice. And the discovery was made that "the North Pole once had a balmy, sub-tropical sea because of extreme global warming," according to the scientists who have carried out this world's deepest drilling into ancient sediment on the far north. Cores retrieved from up to 430 meters below the seafloor in waters 1,300 meters deep show that, for a brief period which occurred around 55 million years ago, the Arctic Ocean was around plus 20 C, compared with today's typical average temperature of minus 2.5 C. The evidence of this huge climate change comes in the form of fossilized marine plants and animals which died out rapidly within a relatively short period of time because they could not cope with the surge in temperatures, while the cores also revealed many fossilized extinct algae that only lived in sub-tropical conditions. Now the scientists are absolutely confident in their conclusion about the massive temperature swing 55 million years ago. As for what might have caused the enormous warming during this period, many theories are being aired but they all have to do with an increased solar activity.

According to University of New England research the Earth is closing to 500-year peak of intense solar maximum activity, which would lead to the dramatic warming of the planet for the next 50 years. These 1500-year and 500-year cycles have been associated with rapid changes in temperature of up to 3 degree of Celsius and ocean level changes of up to 2 meters. This research proves that the ice ages and warming periods recorded in history are linked to the solar cycle. For example, Sydney and Brazil fossils evidence shows 3000 years ago a couple meters sea level rise and very warm period presence - much warmer then now and 900 years ago during the Vikings Medieval maximum of solar activity as well.

At over 1.4 million kilometers or 869,919 miles wide our Sun contains 99.9 percent of the mass of the entire solar system - much over a million Earths could fit "inside" the star. The total energy radiated by the Sun is almost 400 billion trillion kilowatts. It equivalents to the energy generated by 100 billion tons of TNT exploding every second. Time to time (usually during the 11-year solar maximums but exactly when scientists cannot predict) an event occurs on the surface of the Sun that releases a tremendous amount of energy in the form of a solar flare or a coronal mass ejection (well know as a sun spot): an explosive burst of super hot electrified gases with a mass that is much more then the mass of Mount Everest. These Solar flares release tremendous amounts of energy, equivalent to a million hundred-megaton nuclear explosions, in just a few seconds. Till 1859 these flares the problems were important from the astronomical point of view only. However, these huge coronal mass ejections begin to occur with steadily increasing intensity and the Sun activity itself increases with each 11-year cycle.

The 22nd Sun cycle substituted the 21st cycle as the most powerful 11-year cycle with a rated X20 huge solar flare toward our planet in 1989, which damaged transformers of the Hydro-Quebec electric system, leaving over 6 million people in Canada and in the United States without power and it also knocked satellites out of orbits and disrupted most of radio and satellite communications on the Earth. During the current 23rd solar cycle, in year 2001 the strongest ever on the record X21 coronal mass ejection occurred. Likely for us, it missed the Earth, and scientists call it "the largest recorded event... but these flares are not growing bigger." Until October 23, 2003, when something tremendous occurred: again likely for us colossal X40 plus solar flare, much bigger in size then a Sun itself, has risen from the opposite side of the Sun. Although, it was from the opposite side, in knocked out many satellites, some were claimed by Japan as non functional anymore, and set a radio blackout on the entire Earth for several hours. The consequences could be thousands times stronger, if this X40 flare hit toward the Earth like 1989 flare.

Even now, during the deepest pick of the current solar minimum cycle and its decline we observed several whooping solar flares in September 2005. These eruptions were measured at a magnitude of X-20 to X-29 and disabled satellites and an instrument aboard a Mars-bound spacecraft. Earth was spared wider disruption as the flare caught a glancing blow. These recent solar flares have also triggered spectacular auroras borealis, or northern lights, which occur as streams of charged particles from the Sun excite molecules in the atmosphere to produce bright colors in the night sky. They are commonly observed near Earth's north pole and south pole. The only difference in the fact that many people in Arizona and New Mexico were enjoying these amazing sky events during the solar minimum in September 2005. Yes, aurora borealis took place over Southern US!

The estimated time of next solar maximum is year 2011. Aren't you frightened yet by the pervious two cycles and the idea that the new 24th solar maximum predicted would be even stronger then the previous two? At least an astrophysicist from Holland Doctor Van der Meer and some other scientists are in panic - after observation of such abnormal activity they said that our Sun would blow within next six years like a super nova. Hopefully, their theory is an incorrect one, because before the most observed super nova stars explosions their brightness have increased dramatically. The brightness of the Sun is a constant for now (at least according to the recent satellite observations, covering about 20 years the brightness of the Sun increased only by 0.01 percent, while some other stars in our galaxy shows their brightness changes (increasing or decreasing) from 0.5 to 1 percents - they obviously heating and cooling their nearest planets, but certainly they are not going to burst right away). In fact, the amount of light from the Sun penetrating the Earth's atmosphere is even little lower than hundreds years ago, because our industry is creating more tiny particles, called atmospheric pollution, and dust than during the last millennium. These are flying in our atmosphere, creating a modern day Krakatoa volcano effect, but sufficient only to cause an asthma in urban areas and not sufficient to offer any protection from an increasing Sun activity.

In fact, Earth getting darker: its surface receives 10% less of sunshine than 200 years ago mostly due to atmospheric pollution. However, this atmospheric pollution only aim to Earth overheating but creation well known "green house effect." Because of the increasing solar activity and atmospheric pollution planet's oceans overheat as well, creating the super powerful hurricanes like Katrina, which caused over 150 billion in damage and claimed over 1,000 lives in four US states. Such hurricanes are only became possible when ocean surface overheats over 80F - that's the only condition needed for a hurricane to start.

On the other hand, we have nothing more to contradict to Van der Meer's supporters, since no astronomy or physics law is preventing (even theoretically) the Sun from the potential explosion much earlier then 5 billion years AD. These ESA (European Space Agency) scientists are telling us that a temperature of the Sun has raised. This fact, partially proven by a Russian satellite, which equipment allows astronomers to measure the exact temperate of the surface of the Sun which beat all pervious estimates (1-2 million degrees) and determined the temperature was over 2 million degrees with zones of over 10 million degrees of Celsius. However, the temperature inside but not outside the star is really matter, but it cannot be exactly determined using any existing equipment. In other words, according to Van der Meer, the internal Sun temperature has risen critically and uncontrolled thermonuclear reaction has already began - thought to be an exactly the same process which detonated the Copernican star. That explosion of far away super nova was visible on the Earth's sky in 1604 even during daytime. Although the explosion itself happened millions years ago but the light from the Copernican star traveled that entire time trough the vast galaxy down to the Earth. Thus, the light from the Sun travels little more then 8 minutes to our planet. So in case, the Sun will indeed explode tomorrow, we will have the last eight minutes to enjoy the life.

Once again: Van der Meer's theory is (hopefully) incorrect and the Sun will not became a new super nova but just silently dies in 5 billion AD as a huge red giant. Let's give a word to more conservative scientists now.

Dr Robert Baker, is vice-chair of the International Geographical Unions Commission on Modeling Geographical Systems, claimed that evidence born from a computer model of the Sun activity over the past 100,000 years shows direct correlations with solar activity and changes to the climate on the Earth, but "What is alarming is that all these switching events over the past 1500 years have occurred in as little as the last seven years." Ilya Usoskin and his team from the University of Oulu in Finland and the Max Planck Institute for Aeronomy in Katlenburg-Lindau, Germany, also have found that there have been more sunspots since the 1940s than for the past 1150 years.

Indeed, alarming, but from my points of view all this data means only an evidence that we are closing not to a doomsday of the Solar system but to the scheduled peak of a long cycle of Sun maximal activity with more powerful solar flares during the next short 11-year cycle and more solar storms and higher temperature rising on the Earth.

The next peaks of the nearest three 11-year cycles are predicted to be in 2011, 2022, and 2033. There will not be a doomsday for the entire Solar system but almost certainly would a real Armageddon for the human kind on the Earth. Let"s go back to 1859 to understand why.

Even 145 years ago, many people realized something colossal had just occurred. September 1st 1859, within hours, telegraph wires in both the United States and Europe spontaneously shorted out, causing hundreds fires, while the Aurora Borealis (Northern Lights), solar-induced phenomena occurring in regions near Earth"s North Pole, were observed as far south as Rome, Havana and Hawaii, with similar effects at the Earth"s South Pole. What happened in 1859 was a huge solar flare X40 or so in size that hit toward our planet. "What transpired during the dog days of summer 1859, across the 150 million-kilometer (about 93 million-mile) chasm of interplanetary space that separates the Sun and Earth, was this: on August 28, solar observers noted the development of several sunspots on the Sun"s surface. Sunspots are localized regions of extremely intense magnetic fields. These magnetic fields intertwine, and the resulting magnetic energy can generate a sudden, violent release of energy called a solar flare. From August 28 to September 2 several solar flares were observed. Then, on September 1st, the Sun released a mammoth solar flare. For almost an entire minute the amount of sunlight the Sun produced at the region of the flare actually doubled. With the flare came this explosive release of a massive cloud of magnetically charged plasma called a coronal mass ejection. An average Earth direction flare does usually take three to five days to get here. This one took all of 17 hours and 40 minutes," says Bruce Tsurutani, a plasma physicist at NASA"s Jet Propulsion Laboratory.

Earth magnetosphere (ionosphere) is the only natural protection we have from solar flare charged particles penetrating into the Earth atmosphere. When these charged particles from the Sun enter our planet's upper atmosphere, they create shimmering curtains of colored light, known as auroras in the polar night sky. The coronal mass ejections of 1859 and 1989, and overwhelmed and overloaded Earth"s own magnetic field, allowing high-energy particles to penetrate into Earth"s upper atmosphere and caused following numerous problems in last few decades, when the Earth is more developed then in 1859 and the Sun is also more active.

Solar storms reach Earth"s ionosphere, causing disruption of radio communications, navigation and radars systems on ships, aircrafts, and military radars as well. Surges and shortcuts in long electricity transmission lines may cause power and widespread blackouts and brownouts. Damage to microchips and electrical discharges lead satellites to stop operating, causing disruption of telephone, wireless internet, TV, GPS, data communication services, etc. Radiation levels become hazardous to astronauts and passengers of high flying aircrafts. Charged particles hitting Earth"s upper atmosphere are destroying the ozone layer, which protects us from harmful UV ultraviolet radiation. For example X10 solar flare in 1994 caused major malfunctions to two communications satellites, disrupting newspapers, network television and nationwide radio service throughout Canada and resulted damages and loss in revenue estimated to be in the tens of millions of dollars. Other smaller solar storms have affected electric systems ranging from cell phone service and TV signals to GPS systems and electrical power grids.

Furthermore, according to the Russian medics research from the Central Clinic of Transportation Ministry the number of heart attacks increased three times and the number of strokes increased almost 90 percents during solar storms. That is not unexpected, since our heart is also can be considered as an electric organ, which works due to K - Na ions balanced constant cycle of polarizations, depolarizations, and repolarizations. Sun storms do not make exceptions for any electrical devices including our heart electrical conduction system. And that means that thousands of people with various heart arrhythmias and illnesses will die at the moment the next powerful or even medium sized solar flare will hit the Earth atmosphere.

What is going to happen when during the next solar maxim of 2011, 2022, or 2033 the solar flare of X40 or even bigger like in 1859 or 2003 (we were lucky that the last one was from the opposite side if the Sun) will hit the Earth again? A new Stone Age. Literally. By the same token as in 1859 and 1989 it will reach and overload Earth ionosphere with sixteen to twenty four hours after the huge coronal mass ejection occurred. First of all, it will forever shut down all satellites, space stations, and spacecrafts by damaging their microchips and other electric systems - obviously all wireless phones, internet, TVs, radars, GPS, military, and data communication systems will be shut down as well. With high-energy particles deeper penetration into the Earth atmosphere most flying aircrafts and helicopters radars and other electronics will be disoriented, turned off, or burned. Pilots and passengers will be exposed to high levels or radiation. Aircrafts electronic malfunctions will cause hundreds of plane crashes in less then an hour. Then charges particles wave will hit the Earth surface, causing spontaneous wires short outs like in 1989. However, in 1859 all wires on the Earth were a less then 15 years old telegraph. But now we are literally wired out with billions of electric wires so there will be not hundreds of fires like in 1859 but millions of them. The Earth surface will turn into a one shortcutting and burning fireball within several minutes causing the one huge blackout on the entire planet after, most of the electric systems will be forever gone. All people whose life depends on devices such as electric cardio stimulators and other electric life support equipment will eventually die and, since, our heart is an electric organ (as mentioned above) several millions of people with heart arrhythmias and illnesses will be killed as well. The similar situation happened before in 1859 and many times earlier but nor people were so dependable from the electronic technologies neither these technologies had an ability to create so much harm and chaos back then. In 1859 the invention of the telegraph was only 15 years old and society"s electrical framework was truly in its infancy.

According to the calculations, more fully represented in my original Astrophysical research for Ravenhurst University in Amsterdam, Netherlands, and based on the recent increasing Sun activity, and number of large over X10 solar flares during the past few decades such a worst case scenario is pending within next 1 to 4 solar cycles. The chances are 35 percent by 2011 (or during the max of the 24th Sun cycle), 65 percent by 2022 (or during the max of the 25th Sun cycle), 95 by 2033 (or during the max of the 26th Sun cycle), and nearly 100 percent by 2044 (or during the max of the 27th Sun cycle).

However, the mentioned catastrophic event is, actually, a pending soft case scenario of the Sun's behavior. Here is the worst case scenario which is unlikely, but still cannot be completely ruled out during the next 40 years or so of maximal solar activity. Giant solar flare toward the Earth could generate jets of energy or blobs of solar stuff at nearly light-speed gamma ray burst, which would be as bright as the Sun although not in visible light, but in gamma rays. Luckily for us, most gamma rays (nearly all but the highest energy versions) cannot penetrate Earth's atmosphere. But the visible light does. Something that intensive would create an optical flash by scattering electrons in the upper atmosphere and creating something like a super aurora upon the planet. The flash of heat and light capable of flash-burning anything which not in the shade. While heating the atmosphere could generate hurricane intense winds upon the whole Earth, causing the temporary elevation of the Earth's medium temperature for 5-10 C or more. Only this would affect the whole life of the Earth eventually. Furthermore, researchers from the University of California, Santa Cruz state "A massive gamma ray burst could lead to mass extinction on Earth. Gamma rays interacting in the Earth's atmosphere would burn away the ozone layer, allowing deadly ultraviolet radiation to penetrate through the atmosphere. The influx of radiation would lead to widespread cancer and other diseases. ... When these bursts of energy interact with our atmosphere, they would produce a lethal dose of byproducts - particles called muons. Most of the species on Earth - on the ground, underground and in the oceans, seas and lakes down to tens of yards (meters) - will be extinct directly by these penetrating muons." Here are details of the five worst mass extinctions in Earth's history by the Smithsonian Institution's National Museum of Natural History: "1. Cretaceous-Tertiary extinction, about 65 million years ago. The extinction killed 16 percent of marine families, 50 percent of marine genera (the classification above species) and 18 percent of land vertebrate families, including the dinosaurs. 2. End Triassic extinction, roughly 199 million to 214 million years ago. The death toll: 22 percent of marine families, 52 percent of marine genera. Almost all Vertebrates died out. 3. Permian-Triassic extinction, about 251 million years ago. The Permian-Triassic catastrophe was Earth's worst mass extinction, killing 95 percent of all species, 55 percent of marine families, 85 percent of marine genera and an estimated 70 percent of land species such as plants, insects and vertebrate animals. 4. Late Devonian extinction, about 364 million years ago. It killed 25 percent of marine families and 60 percent of marine genera. There were few land organisms at the time. 5. Ordovician-Silurian extinction, about 439 million years ago. The toll: 25 percent of marine families and 60 percent of marine genera." We are probably not on the edge of the sixth massive extinction but we cannot completely rule out such possibility as well.

Now, the situation is aggravated with the fact the Earth magnetic shield that protects us from the solar flares gets weaker every day. Originally, the powerful enough to protect the Earth from the Sun our magnetic shield (ionosphere) is created by the difference of speed rotation of the planet and its inner core. Earth"s outer core rotates in the opposite direction of its inner core, generating a magnetic field is carried by this vortex. However, our planet is slowing down because of some still not researched enough gravity effects. Thus, a billion years ago a day on the Earth was only 18 hours long. In the distant future about 5 billion years from now, a day will be 960 hours long. Because of faster rotation and hotter inner core billions years ago the Earth carry several times stronger magnetosphere, which protected our planet from killing solar radiation and made a complex life possible. In 4 billion AD or so the Earth will completely loose its magnetic shield like the Mars, which also possibly had a life in the past but its extinct now. When the Mars core cooled and the planet slowed down its lost an ionosphere, and a protection from the radioactive Sun. Although, the Mars is much further from the Sun then our planet it became vulnerable to each solar flare in its direction. That is the main reason of Martian satellite and rovers frequent failures. That is why so many missions to the red planet and to some other parts of inner Solar system somehow went wrong, except the last one, when NASA scientist choose year 2004 - the year of the minimal solar activity in the 11-year cycle and not smoothly but still successfully landed two rovers. As you can predict the main problem of human missions to the Mars will be an increased life threatening radiation and vulnerability to the charged particles from the Sun.

However, the main reason why the Earth is loosing its magnetic protection so fast now is not cooling magma or slowing rotation. NASA researcher Bradford Clement casts light on this fact by analyzing records taken from sedimentary samples drilled from various parts the world. He writes: "These samples, deposited at four different ages in Earth"s history, have a residual magnetic echo from the magnetic field that prevailed at the time. These records yield an average estimate of about 7,000 years for the time it takes for the directional change to occur. The big switchover does not take place in one swoop, though. It happens faster at the Equator and takes longer at higher latitudes -- the closer one gets to the poles. The reason for this is that in the absence of the main North-South magnetic field, the Earth"s core develops a weaker secondary field, which has many mini-poles. Eventually the two main poles are established again, but on opposite sides of the planet, and restore their primacy. No one knows what would happen to life on Earth if the "flip" occurred today but the speculation borders on the doomsday. Many aspects of life today would be literally turned upside down, both for humans, given our dependence on magnets for navigation, and for migrating animals which use an inner compass. We would also be more exposed to deadly busts of solar radiation, from which we are normally protected by Earth"s magnetic field. And the loss of that shield would cause solar particles to smash into the upper atmosphere, warming it and potentially causing wrenching climate change."

French geophysicist Gauthier Hulot and US scientist Ronald Merrill also discovered a weakening of Earth"s magnetic field near the poles, which could be a warning that a "flip" is near. However, polarity reversals occur randomly in time. The shortest interval between "flips" is between several thousands years, and the longest is as long as 50 million years. The Earth magnetic field is weakening, and it is a certain sign that the flip is near, but nobody knows for sure when will it happen. Scientific estimates show a time interval between tomorrow and another 30,000 years.

But let us turn back to the other "not correlated" from the fist look event, which happened about 3.5 billon years ago. Let's go to the Mars. Now almost all of the astronomers believe Mars once had seas and oceans of surface water, enough to support life. Many of them believe then the Red planet had a plenty of life as well. However, they have not determined where that water and possibly life itself went some 3.5 billion years ago.

Thus, scientists monitoring after effects of the mentioned monster solar storm that hit the Earth in October and November 2003 so they think repeated buffeting by this kind of space weather could have ripped away Mars' watery veil.

Around 3.5 billion years ago Mars core cooled and lost its magnetic field exactly the same way as the Earth will in the more distant future. These solar radiation events can affect the Red Planet surface because Mars lost most of its electromagnetic protection, unlike our Earth, which still has a protective magnetosphere that guards the planet against bombardment by high-energy particles during almost every solar storm, Mars now has only isolated small and medium size zones of protection. Observations by the Mars rovers Spirit and Opportunity have bolstered the idea of plentiful Martian water, according to Thomas Zurbuchen of the University of Michigan. "Where did it go?" Zurbuchen asked about Mars water. "One of the key ideas that people are talking about is the connection to these space storms ... Over 3.5 billion years, there's kind of a gradual erosion of this water." The brief video, at the NASA Web site, showed water literally blowing away from the planet during the powerful solar storm which took place some 3.5 billion years ago. There is a high probability that Sun whipped out water and life from Mars when its magnetic field was possibly still stronger then the Earth's magnetosphere will be during the mentioned earlier next "poles flip" which will occur between tomorrow and another 30,000 years. Is there are coincided "poles flips" and powerful solar storms wiped out most of the Earth life mentioned five times of mass extinctions or is it asteroids, comets, super volcanoes, giant tsunami, or the Sun alone? No one can give you an exact answer but so far most of the scientific evidences are pointing to the Sun.

Scientists worked with many of robotic spacecrafts to watch the impact of 2003 "Halloween" solar storm, the most powerful ever monitored. They started with the SOHO spacecraft which monitors the Sun itself. SOHO travels a wide elliptical orbit 900,000 miles away from Earth as it monitors electromagnetic ejections from the Sun and provides early warning when solar storms are about to hit Earth. This is the only protection from the Sun that we have so far. Thanks to the 2003 super storm it was partially shut down when its high-gain antenna froze up reducing its ability to transmit data to Earth and its cameras and other science instruments were shut down as well. SOHO was partially restored a few month later. Then the astronomers also tracked the solar blast wave with the Ulysses space craft near Jupiter and the Cassini space craft that is orbiting Saturn. They followed the wave all the way trough the solar system, where the two relatively ancient Voyager probes, launched in 1977, are aiming for interstellar trip. Even some aircrafts on Earth had technical problems as well. In space, astronauts aboard the ISS (International Space Station) had to move into the Russian Service Module, which offering much better shielding from solar storms.

"Solar storms, like a big one that affected Earth last year, might have torn away the water that used to cover parts of Mars. These solar radiation events can affect the surface entire Mars," said Ed Stone of NASA's Jet Propulsion Laboratory in Pasadena, California.

Ancient Egyptians, Greeks, Romans, Incas, Arabs, and many other civilizations intuitively realized the power of our star and were worshipping to their Sun Gods: Apollo, Ra, Tonatiuh, Asuramaya, etc. Today, we are so blinded by our technological development progress that we have forgotten about the main source of life and death in our Solar system - the Sun. One day the Sun made life in our Solar system possible, the other day the star will take it back. At least, the life that we know.

Part Three. Silentio Universum.

Our ancestors many times ask themselves: "Why our universe is silent?" And you probably also ask yourself gazing at the starry sky: "Are we alone?" But hear only a "Silentio Universum" - Silence of the Universe (translation from the Latin expression) instead of the answer.

However, in 1960s the radio technological break through allowed scientists to start one of the most ambitious and expensive projects of interstellar communication - the Search for Extraterrestrial Intelligence - SETI, as it is well known. It made its beginnings on April 8, 1960 under the name "Project Ozma." Today, SETI has become powerful research organization committed to the search for intelligent life beyond the Earth. In last 45 years, SETI's astronomical observing equipment has grown in power by a factor of 100 trillion. SETI programs using most powerful existing radio telescopes and receiving systems to defect weakest radio signals from other stars and galaxies, like those which are coming from the Earth to them. If there are any developed alien civilizations present, they almost certainty would use a radio-like communication, so SETI radio telescopes would detect those signals for sure. SETI-like institutes were founded in Russia, Australia, China, Spain, the United Kingdom and elsewhere in Europe as well. However, they all failed to find any evidence of intelligent life beyond our planet.

US SETI program is still alive and developing because of private funding although US congress had stopped financing this project in early 1990s, since most of the scientist thought that our Sun has unique planetary system, which made life on the Earth possible, while the other stars do not have any planetary systems at all. However, the fist planet orbiting another star was discovered in 1995. Since this first planet discovery, many hundreds extra solar planets have been found. Although most of these planets are nothing like our solar system inner planets: they are huge - mostly many times the size of Jupiter - and some are more like another class of object, the failed stars - brown dwarfs - the stars which unlike our Sun failed to ignite thermonuclear reactions inside them due to small size and mass. But in 2004 several Earth-like planets were found as well.

As scientists can predict hundreds of billions of planet are present in our universe, some of them are Earth-like and some of them are certainly could support life in our understanding of this word. Nevertheless all SETI-like institutes and the other life-search programs have failed to catch anything for 45 years.

Here is why. The conditions needed for complex life existence are very rare in the Universe. Looks like many newly found Earth-like planets are located in our galaxy's spiral arms, where they are subjected to frequent super nova explosions, black holes, and comets impacts. Stars like our Sun are also become monstrous fireballs by the end of their life cycles and swallowing their planets. As we learned in the pervious chapter, planets without its own magnetic filed are also unable to support not even a complex life but even relatively simple forms of life. Most of old planets, which had enough time to develop a complex life like our Earth, are cooled and lost their ionosphere protection from stars prominent UV radiation. If a planet is located little closer to its star it is a liquid boiling ball of magma, if a little further - then it is a frozen ice ball, and both such systems are unable to support any kind of life from the modern scientific point of view. Thus, the distance from the center of Earth to its surface is 3,963 miles (6,378 kilometers). Much of Earth is a liquid magma and only its surface has a unique ability to support our life. The more or less solid shell of the planet is only 41 miles (66 kilometers) thick. Relatively speaking, it is much thinner than the skin of an apple.

Satellite observations by Lockwood and Skiff of 40 nearby stars similar in size and makeup to our Sun, revealed that a half of them varied by 0.5 - 1 percent in brightness in less than 4 years. Had such 1 percent variations occurred with the Sun, it would be a doomsday for the most of the Earth leaving creatures. It is likely that some of such pulsating stars have large undiscovered planets orbiting them, but the majority of such stars brightness variation has nothing to do with any orbiting planets but depends from the thermonuclear reaction rate in the star itself.

A star is the main component of life in any solar system. But there are less and less stars are remains in our Universe. According to the computer modeled system designed in Pennsylvania University of 100 thousand galaxies in our Universe birth of new stars occurring several times slower then death of old stars. The peak of the birth of new stars was 5 billion years ago. And between the Big Band 14 billion years ago and birth of our Sun 4.7 billion years ago half of the present stars were already created. Galaxies have billions of stars less now then several billions years ago. In other words, you could see much more stars standing on the newborn Earth then now. Space is much more dark then several billions years back.

Thus, if any alien civilization has stuffiest time and ability to develop their technology to survive these harsh conditions (which is possible statistically but not much practically), they are probably escape their planet and floating in the deep space in an aggressive search for a new planet with needed recourses. Our Earth could be a perfect place to live not only for terrestrial creatures. So the practically realized idea of sending the constant radio signals to the outer space in order to notify all who can hear about our presence may be as not good as it seems to be. Try to make notice in a jungle and you will be eaten fast. But, hopefully, we were unbelievable lucky to survive in space, so we are yelling in a desert but not in a jungle. And it is more useless then dangerous.

So what about all those UFOs, mysterious signs of the fields, area 51, and all other aliens stuff? In fact, 99.9 percent of "UFOs" are actually flying objects such as satellites, meteozonds, various flying military and scientific devices, aircrafts, and rare atmospheric events (such as thuderballs, coronas, light refractions, etc). The other 0.1 percent of UFOs and mysterious signs of the grass and raw fields are deliberately created. How and why? The movie industry (both fiction and non-fiction) and newsmakers and benefiting the most from nearly each confirmed case of UFO seeing or signs appearing. No UFO means no sensation, no sensation means no money. I have a little doubt that some UFO-oriented corporation or a group of corporations launched a secret project to make aliens real even if they never existed. And our imagination works well to help them. The cost of making a radio controlled well build UFO is far less then $5,000 while the cost of the direct and not direct profit from it could be much more then $50,000,000.

Furthermore, Russian scientists have proved that to create needed signs on almost any field of plants such a grass, raw, etc, you will a powerful microwave capable to focus and beam its waves from several kilometers. The cost of such special equipment and its installation on any communications, weather, gps, spy or other satellite would cost you several millions. But now compare it to the dozens of millions with certain filmmakers, book writes, newsmakers, and even scientists made researching this "alien" mystery.

All of this are the kinder garden plays so far; however, let us ask "Why does US government keeps zone 51 and several other "alien secret objects" closed fueling interest to these "secret of the secrets?" Why don't they just open some of them and show that there are actually nothing more than several rusted UFO-like pieces of old experimental aircrafts?   The day when bio scientists can generate or clone proper controllable alien-like super humans which not only extraordinary thinking capabilities is just a couple dozen years from now or even less. These would be the days of depleted Earth when not much of natural recourses left, and the entire planet will pass trough catastrophic events of global climate changes, overpopulation, pollution, abnormal Sun activity, sharp energy shortage, wars and instability, when the global economic and political collapse is unavoidable. According to the non secret Pentagon report the beginning of this period of "global instability" is already started. Thus, during the mentioned events only the miracle would save the world from the global economical collapse and political chaos, so it seems much possible that a while ago top officials had realized the unavoidable course of the future and some researching facilities had already stated to create that "miracle" for the future days. It is a quite possible, that one dark day governments would have no choice to hold power and maintain control but to tell "THEY finally here to save us" transferring the part of the functions to the newly cloned extra smart and well controllable creatures. This is not so fantastic after all; otherwise, why someone so desperately needs aliens to exist even today?

Does it mean we never will be able to find any kind extraterrestrial life? No, in fact, we already have found it. Or better to say, they found us. Estimates are different, but the USGS says that from 1,000 million grams to 1,000 tons of material enters the atmosphere each year to Earth's surface. And there are only the simplest forms of life proven by lab experiments to survive radiation and cold are tiny prion-like viruses and some bacteria. It is hard to believe that many maximums of solar activity coincide with all major influenza pandemics (world wide epidemics).

Increasingly frequent solar flares and solar wind are driving these organics virus triggers out of the stratosphere and into your body, where an inactive influenza virus is already present, to activate it. Yes, there are tiny organic germs from outer space, which are deposited throughout space by dust, and the debris such as stream of comets and meteors, which could harbor organic material as well. As the Earth flying through the stream or dust that contain virus triggers, these prion-like organisms enters our atmosphere, where they can stay for some time, until gravity pulls it down but more frequently a Sun coronal mass ejection toward the Earth driving these viral particles rapidly to the ground level causing world wide epidemics.

There were several strong evidences that human contact cannot be accounted as the only reason of the influenza spread. In 1918, an influenza outbreak occurred on the same day in Bombay (India) and Boston (North America), yet that epidemic took another whole three weeks (22 days) to reach to New York, which is located only several dozen of miles away from Boston. Also there were numerous cases when people or groups of people separated from others mostly on sailors' ships were catching influenza without contacting others for several months during an epidemic outbreak.

Cardiff University researcher Chandra Wickramasinghe states that the deadly disease SARS, or severe acute respiratory syndrome, might have come from outer space as well.

"Over time, gravity performs a few plausible, but not routine tricks, and the comet is ejected from its stellar orbit like a rock from a slingshot. For more than a 100 million years it slips silently across the inky vastness of interstellar space. Then gravity goes to work again. Another star tugs at the comet, pulls it in. A few giant gaseous planets whiz by, their bulks tugging at the comet, altering its course slightly. Ahead now, growing larger, looms a gorgeous blue and brown marble. Water and land. Maybe some air. Then with the force only the cosmos can summon, the comet slams into the third rock from a mid-sized, moderately powerful star. The alien microbe survives, emerges from its protective shell and spreads like the dickens" - this is how life on Earth has began 3.8 billion years ago, according to the theory called panspermia, which states that seeds of life is everywhere and that we humans owe our genesis and evolution to a continual rain of foreign microbes, which means, that we might all be aliens after all.

The pansermia theory has four main postulates: "1. Life got started on a cosmological scale including the combined resources of all the comets around all the stars in all the galaxies of the entire universe. 2. The space between stars is littered with cometary debris, some of which contains the seeds of life. 3. Comets arriving at the Earth from the 100 billion-strong Oort cometary cloud of our solar system brought the first life onto our planet some 3,600 - 4,000 million years ago. 4. Evolution of life on the Earth was directed by the continuous arrival of outer space micro organisms which must still be arriving at the present time."

As we know all attempts to create life from non life, starting from Doctor Pasteur, have been unsuccessful. Recent geological evidence indicates that life was present on the Earth over 3.6 billion years ago, at a same time as the planet was pummeled by comet and meteorite impacts. Of course, not all microbes in interstellar space would survive, but the survival of even the smallest fraction of microbes leaving one solar system and reaching the next site of planet formation would be enough for panspermia to be one hundred percent more probable than starting life from scratch in a new location as we were taught all this time by leading biologists.

The Aerospace Corporation in Los Angeles study reports on data collected during the 1998 and 1999 Leonid meteor showers: "The annual event, which peaks again this weekend, occurs when Earth moves through a stream of debris left behind by comet Tempel-Tuttle. That comet passes through the inner solar system every 33 years, with its grains of dust zipping along at 160,000 miles per hour (72 kilometers per second) relative to Earth. When they hit our atmosphere, friction vaporizes many of them. From the ground, we see blazes of light commonly called shooting stars. But studying small meteors from the ground can be frustrating. So Jenniskens and his colleagues at the Aerospace Corporation in Los Angeles, along with other researchers, used two airplanes to create "stereoscopic" images of the meteors. Ground-based instruments were used, as well. In one striking image they followed a meteor that exploded into what scientists call a fireball. The trail left by the fireball contained what the researchers called the "fingerprint of complex organic matter. The fingerprint involves much higher-than-expected concentrations of carbon monoxide and carbon compounds that seemed to develop as the meteor interacted with Earth's atmosphere."

The same time researchers While Hoyle and Chandra Wickramasinghe say that the virus, or a trigger that causes it, is deposited throughout space by dust in the debris stream of comets and perhaps meteorites, which are able according to many researchers to harbor organic material. As the Earth passes through the stream, dust (and perhaps the virus or viral like organism) enters our atmosphere, where it can lodge for two decades or more, until gravity pulls it down. The intense Sun activity at sunspot maximum has the effect of driving viral particles or their triggers rapidly to ground level. Ionized gas from solar flares is channeled to the Earth along its magnetic field lines. The flow of charged particles from the Sun generates electrical fields across the planetary stratosphere, accelerating the down flow of virus or its organic triggers. Then, in lower levels of the atmosphere, these particles condense, ultimately coming down in raindrops. That is why there is much higher probability (which is confirmed statistically) that you may get ill under the rain then under the sun. Try to not swallow alien microbes under the rain, during the solar maximum because you may get influenza, or SARS, or even a common cold. Remember that all major influenza and SARS epidemic somehow coincided with the cycles of the Sun maximum activity. Based on this data we can expect new major pandemics around 2011, 2022, 2033. Organic triggers from the other space are the only answer for these epidemics.

As we learned the Universe is a somewhat deadly place for the life. Then how the life could appear and survive for so long? The answer is: only the simplest forms of the life (its' triggers) could survive the harsh conditions of the outer space. The more complex organisms which evolve and evolution on the planet periodically obliterated by some cosmological events, while the life itself cannot be obliterated completely once it evolved in the Universe. It will end only in 15 billions years from now when the gravity of constantly accelerating Universe will tear apart every galaxy, every solar system, every star, every plant, every creature, every molecule, and every atom with its subatomic particles.

Where does the life fist appear? No scientist know the exact answer but here is the most probable scenario according to NASA research: "Supernova stars events are the incendiary deaths of huge stars several times as massive as the Sun. They use up their nuclear fuel quickly, in just a few million years, and gravity pulls remaining material rapidly inward. Upon collapse, a dying star's core rebounds to generate a shock wave that blasts its outer layers into space. For a moment, it can shine as bright as a 100 billion suns. The first stars in the universe were very massive and made almost entirely of hydrogen with some helium. They were the first chemical factories, forging new elements - heavier with each generation - that were cast into space when the stars exploded. Subsequent generations of stars formed from this detritus. More modern stars, like our Sun, are the beneficiaries of stellar evolution, containing an abundance of heavier elements. The very ingredients for life were created in supernova. Supernovas are the source of the heavier elements, the material we are made of. The Sun and its swarm of planets formed about 4.6 billion years ago in an already ancient galaxy due to a supernova death. The heavy elements that make up of rocky planets, as well as the elements necessary for life on (at least) one of those planets, were forged in mighty supernova explosions. Furthermore, there is evidence the gaseous nebula that formed our proto-sun was seeded by a single nearby explosion. Among the elements ejected by supernovae are some that are unstable and subject to radioactive decay, but with very long decay times: uranium 235 (half-life of 0.7 billion years); potassium 40 (half-life of 1.4 billion years); uranium 238 (half-life of 4.5 billion years) and thorium 232 (half-life of 14 billion years). These elements were trapped in the gas that formed the Sun and planets. They were concentrated in the core of Earth by processes of settling and differentiation when the core of the young Earth was molten from the heat of impact of the planetoids from which it formed. Eons since, the impacts have dropped off, but the long-lived unstable elements -- potassium, uranium, thorium -- continue to decay. Those decays, atom by atom, yield the heat that maintains the molten state of Earth's core. The continental plates float on that core, shifting and colliding, energizing the "ring of fire" around the Pacific basin, wherein lie the Puget Sound El Salvador. Other plates collide to lift the Himalayas and leave treacherous ground in India, Turkey and elsewhere."

Our life is a carbon - hydrogen based life. Every chemical substance containing carbon and hydrogen is an organic substance. There is no shortage of hydrogen in space, however until recently carbon was not discovered in space. But now, a team of Spanish astronomers has made the first detection of interstellar rings of carbon, the type of molecules upon which Earth's life is based. According to the team, led by José Cernicharo (Instituto de Estructura de la Materia, CSIC) "the team used the European Space Agency's Infrared Space Observatory (ISO) to find benzene, the ring molecule par excellence. They think benzene is produced by stars at a specific stage of evolution, and that it is an essential chemical step towards the synthesis of more complex organic molecules whose true nature is still unclear. In industry, benzene is obtained from petroleum and has many uses. Benzene is made of six atoms of carbon chained together to form a ring, plus six atoms of hydrogen, one per carbon. Chemists know today that benzene-type molecules make a whole family of compounds, called aromatic hydrocarbons because of their smell (they are basic constituents, for instance, of perfumes and candles). Astronomers expected to find these ringed molecules in space, where long strings of carbon atoms have been detected. Moreover, it had been postulated that certain compounds of yet unknown nature, that are known to be very abundant in space, are actually aromatic hydrocarbons. These compounds have left their chemical fingerprints, called Unidentified Infrared Bands (UIBs), in many places in the universe. The team think in some places there could be a few molecules of benzene per cubic centimeter, a value considered to be high, although the estimated density of molecules of all kinds in the area observed is 10 million per cubic centimeter." They also state that this molecule is the "missing link" between simple carbon molecules observed in red giant stars, which made of no more than eight carbon atoms, and complex molecules responsible for Unidentified Infrared Bands (UIBs), known to be made of hundreds of carbon atoms. "The missing-link idea is based on observations of objects at each stage of evolution. UIBs have been detected around stars that are already "dead" - the planetary nebula - but not in the previous evolutionary stage of protoplanetary nebula, such as CRL618, where benzene has been found. The molecules causing the UIBs must form in the relatively short period from protoplanetary nebula to planetary nebula" It seems that carbon-rich protoplanetary nebulae are the best organic chemistry factories in space.

The same time Preliminary measurements of interstellar dust particles encountered by NASA's Stardust spacecraft indicate the surprising presence of large tar-like molecules that scientists said could have played an important role in sparking life here on Earth. This finding, could imply that interstellar particles constituted an important delivery system for the molecules necessary for life to begin on Earth billions of years ago because when the large tar-like molecules in contact with liquid water on the Earth, they also could have triggered the type of chemical reactions which are prerequisite for the origin of life.

In the 1990s, many scientists accepted the theory of Panspermia created by Hoyle and Wickramasinghe in 1970's which states that life on Earth had come from space by some kind of life spores or life triggers which are distributed throughout the Universe. If so, would not they still be coming?

Appendix I: Sun Structure.

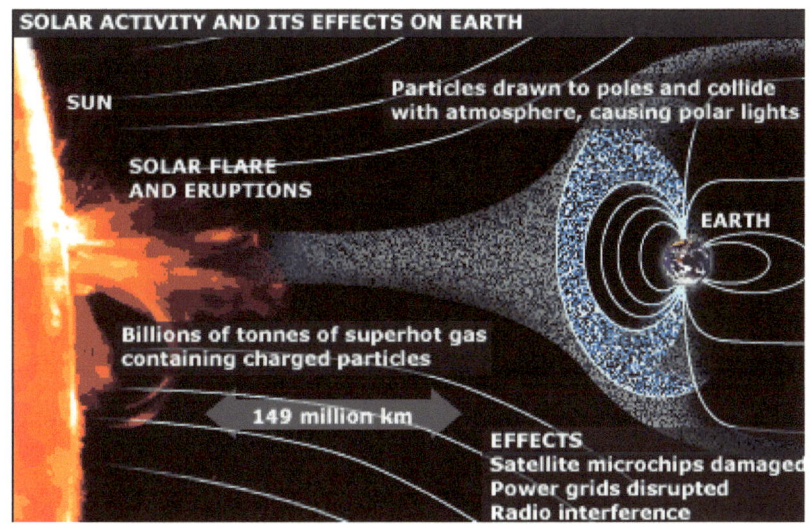

SOLAR ACTIVITY AND ITS EFFECTS ON EARTH

SUN

SOLAR FLARE AND ERUPTIONS

Particles drawn to poles and collide with atmosphere, causing polar lights

EARTH

Billions of tonnes of superhot gas containing charged particles

149 million km

EFFECTS
Satellite microchips damaged
Power grids disrupted
Radio interference

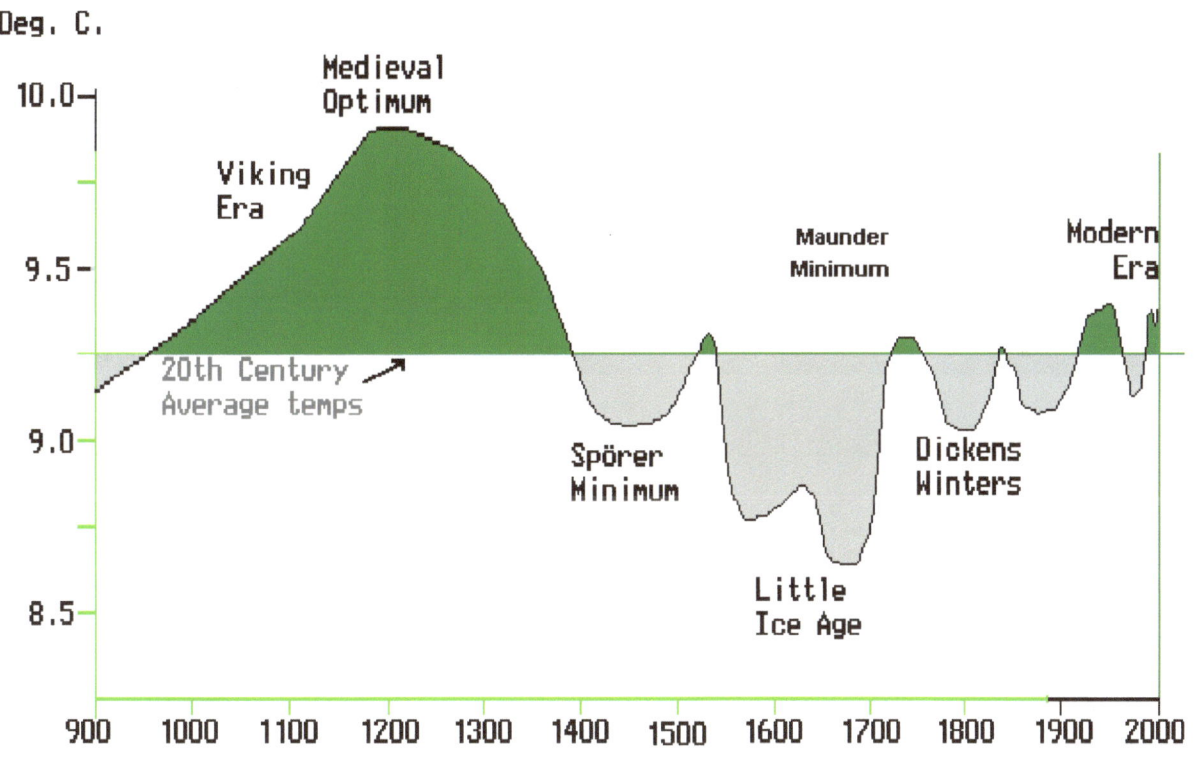

THE LAST THOUSAND YEARS IN EUROPE. Climatic changes.

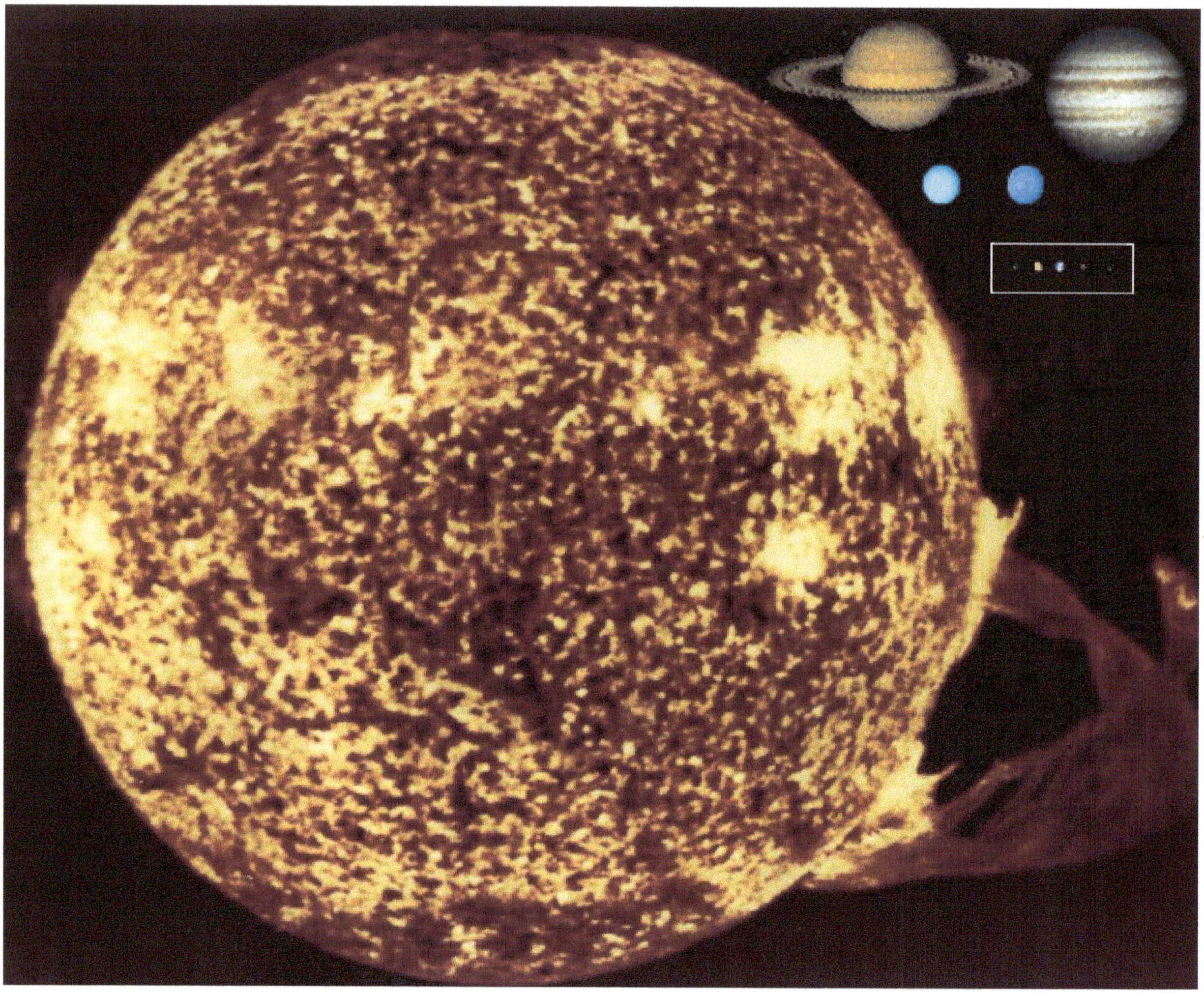

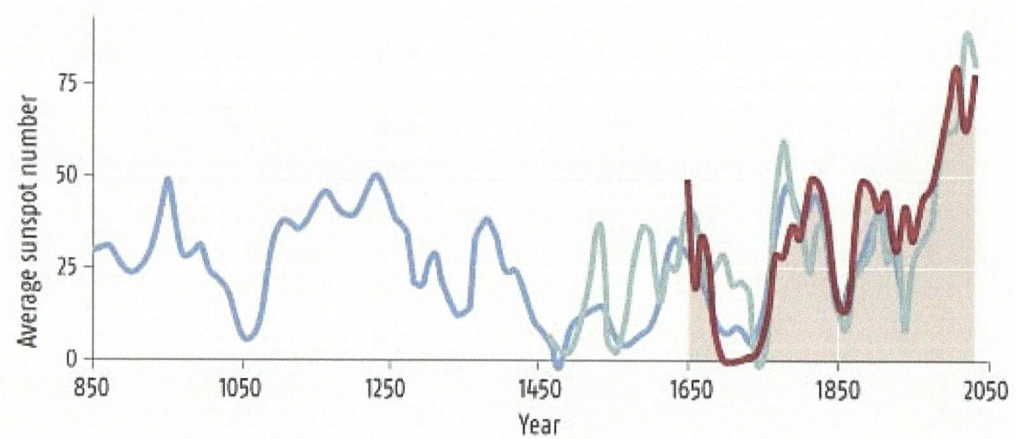

SUNSPOT ACTIVITY

Sunspots are more frequent now than at any time for more than 1000 years

Observed sunspot number
Antarctic readings
Greenland readings
Average sunspot number
0
25
50
75
850
1050
1250
1450
1650
1850
2050
Year

The SUN

The Sun is our nearest star. It is a huge, luminous ball of gas like all the other stars. It consists mostly of hydrogen, and helium with tiny amounts of other elements

The Corona is the outer shell of the Sun's atmosphere. It is extremely hot with temperatures reaching up to 2 million degrees

The Radiative Zone Here energy is transported outwards by radiation. It covers about 70% of the Sun's diameter

The Core In the centre of the Su the energy is produced by fusion processes through which hydroge nuclei are fused to produce helium nuclei

The Chromosphere is a transparent layer above the photophere. It extends up to 2000 km with temperatures around 10,000 K

Sunspots

The Convective Zone It extends roughly over 30% of the Sun's diame Here energy is mainly transported upwards by convective streams of ga

The Photosphere is the visible 'surface' of the Sun. It is about 300 km thick. Here most of the Sun's activitiy takes place, for example sunspots

The Spectrum of the Sun not only shows the rainbow colours: It also shows dark lines named absorption lines or Fraunhofer lines

The solar cycle:sunspots and other forms of solar activity vary with an average period of 11 years

Spectrum of the Su

Granulation

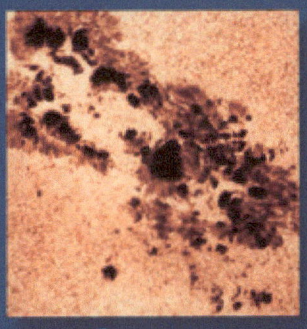

Sunspot

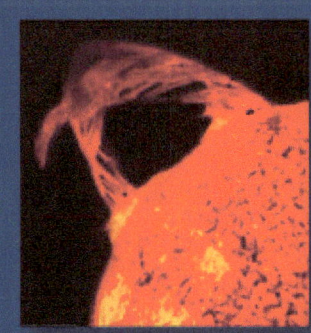

Eruption

The Sun's corona

Appendix II: UV Radiation.

Sunlight UV radiation as a major contributor of skin cancer.

Sunlight has a tremendous effect on the skin causing premature skin aging, host of skin changes, and most importantly, skin cancer. Exposure to ultraviolet light (100 - 400 nm) from sunlight accounts for 95% of three main skin cancers which are "melanoma (MSC), basal cell carcinoma (BCC), and squamous cell carcinoma (SCC)" (1). Although many factors could be involved in pathogenesis of cancers such as sunbeds, tanning salons, wireless radio equipment, low frequency fields of electric power supply, and high frequency electromagnetic fields such as cellar phones and radio telephones; the most important key factor in formation process of these skin cancers is sunlight UV radiation.

Sunburns and excessive exposures cause cumulative damage which induces immunosuppression and skin cancers. Ozone depletion, the level of UV light, elevation, latitude, altitude and weather conditions influence the emission of UV radiation reaching the earth's surface." (7). In the United States there are estimated "7,400 deaths and 40,500 new cases of skin cancer annually. The lifetime incidence of melanoma in the United States is 1 per 90 persons." (4). Even all mighty politics like Bill Clinton or John McCain are not expectations from this skin cancer statistic and were fighting skin cancer as well.

Thus, according to German scientist J Bernhardt statistical research "The exposure to UV radiation must be considered to be far the highest risk. The annual rate of about 2,000 deaths from skin cancer in Germany, mainly caused by extensive exposure to solar UV radiation." (2)

The sun gives off its ultraviolet radiation UV that scientists divide into categories based on the wavelength: UVC - 100 to 290 nm, UVB - 290 to 320, and nm UVA - 320 to 400 nm. The deadliest UVC radiation is almost completely absorbed by the earth magnetosphere and ozone layer and therefore has minimal effect on us. But UVC radiation also can be found in artificial sources such as mercury lamps or germicidal lamps.

The second type - UVB can penetrate Earth atmosphere and affects the outer layer of skin, the epidermis, which is the primary agent responsible for sunburns. It is the most intense from 10:00 am to 2:00 pm when the sunlight is highest above the horizon. It is also more intense during late spring to early fall accounting for over 70% of a person's yearly UVB dose. However, UVB does not penetrate glass.

Finally, the third type UVA is a major contributor to skin damage. UVA penetrates deeper into the skin then UVB. The intensity of UVA radiation is much more constant than UVB without the variations during the daytime and throughout the year. And UVA is also penetrates glass.

According to the other German scientist H Lang extensive research "Both UVA and UVB radiation can cause skin damage including wrinkles, lowered immunity against infection, aging skin disorders, and cancer. However, we still do not fully understand the process. Some of the possible mechanisms for UV skin damage are collagen breakdown, the formation of free radicals, interfering with DNA repair, and inhibiting the immune system. In the dermis UV radiation causes collagen to break down at a higher rate than with just chronologic aging. Sunlight damages collagen fibers and causes the accumulation of abnormal elastin. When this sun-induced elastin accumulates, enzymes called metalloproteinases are produced in large quantities. Normally, metalloproteinases remodel sun-injured skin by manufacturing and reforming collagen. However, this process does not always work well and some of the metalloproteinases actually break down collagen. This results in the formation of disorganized collagen fibers known as solar scars... UV radiation is one of the major creators of free radicals. Free radicals are unstable oxygen molecules that have only one electron instead of two. Because electrons are found in pairs, the molecule must scavenge other molecules for another electron. When the second molecule looses its electron to the first molecule, it must then find another electron repeating the process. This process can damage cell function and alter genetic material. Free radicals cause cancer by changing the genetic material, RNA and DNA, of the cell...UV radiation can affect enzymes that help repair damaged DNA. Studies are being conducted looking into the role a specific enzyme called T4 endonuclease 5 (T4N5) has in repairing DNA... The body has a defense system to attack developing cancer cells. These immune system factors include white blood cells called T lymphocytes and specialized skin cells in the dermis called Langerhans cells. When the skin is exposed to sunlight, certain chemicals are released that suppress these immune factors... The last line of defense of the immune system is a process called apoptosis. Apoptosis is a process of cell-suicide that kills severely damaged cells so they cannot become cancerous. This cell-suicide is seen when you peel after a sunburn. And UV exposure prevent this cell death allowing cells to continue to divide and possibly become cancerous." (3)

As we can see overexposure to sunlight is the major cause and the key factor in skin carcinoma development. Even in such northern country as Germany there are over 2,000 deaths each year due to overexpose to UV radiation, while "in the United States there are estimated 7,400 deaths and 40,500 new cases of skin cancer annually. The lifetime incidence of melanoma in the United States is 1 per 90 persons." (4).

Thus, antioxidants are capable of scavenging reactive oxygen (co-called free radicals), generated during photooxidative stress caused by sunlight. Vitamins C and E together protect our skin from the suppression of the local immune response by UV radiation and itself suppressed the sunburn reaction (erythema). Oral cartenoids are also very effective in diminishing sensitivity to ultraviolet light because they contain both alpha and beta carotene.

However, the most accessible, reliable, and cheap method is regular green tea. "Green tea contains natural antioxidant polyphenolic compounds known as epicatechins. Researchers have shown that green tea polyphenols -- taken orally or applied topically -- exert photoprotective effects that inhibit ultraviolet radiation-induced skin tumors (tumorigenesis). Studies have also shown that green tea extract possesses anti-inflammatory activity, protecting against ultraviolet (UV) light-induced skin inflammation (erythema). The major polyphenolic chemopreventive constituent in green tea responsible for these biochemical or pharmacological effects is epigallocatechin-3-gallate (EGCG)." (6)

Unfortunately, most of the people reduce their intake of these supplements during spring and summer. "While colds and flu may occur less frequently in warmer months, excess exposure to sunlight can impair the immune system and increase risk of cancer." Antioxidants, carotenoids, vitamins and green tea have been shown to shield the skin from UV radiation.. Continued intake of these supplements, "in conjunction with sunscreen and protective clothing, may pay off in the long run with smoother unblemished skin and reduced risk of developing skin cancer." (6)

Considering the fact that "the level of solar activity during the past 70 years is exceptional, and the previous period of equally high activity occurred more than 8,000 years ago. We find that during the past 11,400 years the Sun spent only of the order of 10% of the time at a similarly high level of magnetic activity and almost all of the earlier high-activity periods were shorter than the present episode." (5) Thus, in order not be included in that sad annual statistic scientist recommend "protective measures in the order of their effectiveness are protection by adaptation of behavior, by clothes, sun hats and sunglasses as well as by sun creams...With regard to UV tanning appliances it is recommended not to use artificial UV radiation for cosmetic purposes because of the related health risks. (2)

PubMed Bibliography:

1. Ramos J, Villa J, Ruiz A, Armstrong R, Matta J. UV dose determines key characteristics of nonmelanoma skin cancer. Cancer Epidemiol Biomarkers Prev. 2004 Dec;13(12):2006-11.
2. Bernhardt JH. Electrosmog, cellular phones, sunbeds etc. -- adverse health effects from radiation? Health aspects of non-ionizing radiation. Bundesgesundheitsblatt Gesundheitsforschung Gesundheitsschutz. 2005 Jan;48(1):63-75.
3. Lang H. Molecular mechanisms of the biological effects of UV radiation. Z Gesamte Inn Med. 1976 Dec 1;31(23):959-61.
4. Ries LAG, Hankey BF, Miller BA. Cancer statistics review 1973-1988. Bethesda, Maryland: National Cancer Institute, 1991. NIH Report No. 91-2789.
5. Solanki SK, Usoskin IG, Kromer B, Schussler M, Beer J. Unusual activity of the Sun during recent decades compared to the previous 11,000 years.Nature. 2004 Oct 28;431(7012):1084-7.
6. Fuchs J, Packer L. Antioxidant protection from solar-simulated radiation-induced suppression of contact hypersensitivity to the recall antigen nickel sulfate in human skin. Free Radic Biol Med 1999 Aug;27(3-4):422-7
7. Saladi RN, Persaud AN. The causes of skin cancer: A comprehensive review. Drugs Today (Barc). 2005 Jan;41(1):37-53.

About The Author.

The author Igor Kyan, MS was born in 1979 in Kiev, Ukraine, lives in the Unites States since 1999. Graduated with Honors in National Ukrainian Medical University (AA Degree in Health Science - 1999), San Francisco State University (BS degree in Biology - 2003), Amsterdam Ravenhurst University (MS degree in Cell and Molecular Science - 2004). Except his current astrobiology research performed for Amsterdam Ravenhurst University, Netherlands, the author is also well known for his graphic artworks and previously published books "Graphic Art " in 2001 (over 10,000 hard copies sold) and "Answers" in 2005. The author is also a bearer of many awards in art and science. Igor Kryan's books goal is to provide his readers with understandable answers for complicated questions which remained unanswered till now.

www.ingramcontent.com/pod-product-compliance
Lightning Source LLC
Chambersburg PA
CBHW051102180526
45172CB00002B/740